How to Make a COOPERED WOODEN BUCKET

How to Make a COOPERED WOODEN BUCKET

A Beginner's Guide with
Detailed Instructions
and Illustrations

by
JAMES D. GASTER,
Vintage Cooper

Text by David Gasten

Pleasant Word

Packaged by Pleasant Word, PO Box 428, Enumclaw, WA 98022. The views expressed or implied in this work do not necessarily reflect those of Pleasant Word. The author(s) is ultimately responsible for the design, content and editorial accuracy of this work.

Photos by Marilyn Gaster, except photo on page 13 by David Gasten and back cover photo by Kimberly Zarybnicky.

Drawing credits: All drawings in Chapter 3 by Amy Swanson, cooper's axe head drawing in Chapter 4 by David Gasten, Figure 6–75 and all drawings in Chapter 5 by Mr. Gaster, and all other drawings in Chapter 6 by Jimmy Sommers.

ISBN 1-4141-0137-6
Library of Congress Catalog Card Number: 2004101165

"Do you see a man skilled in his work? He will serve before kings; he will not serve before obscure men."

Proverbs 22:29

Table of Contents

Acknowledgments 9

Chapter 1: Introduction 11
Chapter 2: Types of Coopering 14
Chapter 3: Nomenclature 16
Chapter 4: Tools and Materials Needed 18
Chapter 5: Making Stave Gauges 26
Chapter 6: Making a Wooden Bucket: The Process in Detail 30
1: Selecting Wood 30
2: From Staves to Bucket 36
A: Shaping and Binding the Staves 36
B: Making and Adding Temporary Hoops 46
C: Cutting and Croze 54
D: Assembling and Inserting the Head 59
E: Smoothing the Outside of the Bucket 71
F: Making and Adding Permanent Hoops 73
G: Smoothing the Edges of the Bucket 75
H: Making and Adding the Bail (Rope Handle) 75
I: Finishing 78
Chapter 7: How to Care For Your Bucket 81
Chapter 8: Gallery 83

Sources / Bibliography 87
About the Author 90

Acknowledgements

Like many projects, this book would not have been possible without the help of many others, and I would like to extend my gratitude here to those who helped make this book possible. First I would like to thank my wife Marilyn for laboring alongside me on this project and believing in the impossible. Also to David Gasten for the many hours of hard work that went into transferring my ideas to text and creating the rough layout. Thanks to Jimmy Summers and Amy Swanson for contributing the diagrams. Also, thanks to Tom Kelleher for his suggestions on improving the instructional chapter, to Judy Howard for her many helpful suggestions, to Country Workshops for their advice and assistance, and to Lori Hinze for copy editing. And finally, thanks to The Lord for opening doors that no man can open.

—James D. Gaster

Chapter 1
Introduction

The wooden bucket carries with it a comfortable sense of familiarity, an aura of pleasant nostalgia of a bygone way of life—one of stone farmhouses, red barns, and wishing wells; of Bessie the cow, chickens merrily pecking at the ground, and children collecting milk and eggs. You might think of the American pioneers or the Swiss Alps, but either way, the wooden bucket is a symbol of a way of rural life that has almost completely been retired to books, movies, and the imagination. You will find the wooden bucket lingering in the fringes of almost any film or TV show about country life in the past, from the Middle Ages to the early American West. In addition, the bucket is also a figure in living history museums throughout North America and Europe.

But where is the wooden bucket in our modern-day landscape? It has been replaced with plastic, steel and cardboard, all of which have become inexpensive and easy to manufacture, and have proven themselves more practical and durable than a wooden bucket ever could be. But none of these newer arrivals possess the beauty and gentle character of the wooden bucket. The metal bucket may suffice in an Old West theme, but do you really want to accent those flowers by putting them in a paint can or a five-gallon plastic bucket?

As a cooper and a craftsman, wooden buckets and similar handcrafted items are my livelihood, and my work has appeared in films, on television, and in museums throughout the United States. I have written this book for one reason: that you may be able to experience the bucket-making process firsthand. I will be explaining the process with the cooper's specialized tools but will be simultaneously recommending substitutes whenever possible, so that you will be able to use the tools you have at hand. I will walk you in detail through the process of making my company's most basic bucket design, the Primitive Camp Bucket. I will support this with some basic knowledge of the coopering trade and an explanation of the names of parts of the bucket and processes used, as well as an explanation of the vintage tools in the cooper's toolbox.

From the beginning, be aware that crafting a bucket, a butter churn, or a barrel is more difficult and specialized than many other woodworking skills, which is why these items belonged to a special trade and were not made by carvers or carpenters. Much of the challenge in building a wooden bucket is in the fact that you are piecing together and then binding a free standing, watertight wooden cone, but that challenge is also what will make the project fun.

Throughout the book, your work will be limited to hand tools. Tool collectors and experienced craftsmen will agree that hand tools add a personality and charm, or a "human element", to the finished piece. I recommend making the bucket with hand tools first before experimenting with power tools, because in using hand tools, you as a craftsman will gain a better firsthand knowledge of the craft. As you have heard before, "Experience is the best teacher".

Read on, and I hope that before long you'll be ready to make a bucket of your own.

Chapter 2
Types of Coopering

You may have heard the word "cooper" and just thought of it as a last name, but, as with Smith, Mason, and Miller, it is a reference to an actual trade. A cooper was a specialized form of woodworker whose job was to bind long, thin strips of wood into usable containers, such as barrels, buckets, butter churns, and wash basins. Before the advent of steel, cardboard and plastic containers, wood was the material in common use, and coopering was the profession that provided these containers.

Types of Coopering

There were three types of coopering, and each form required a different amount of skill than the next. They were:

a. Dry coopering
b. Wet coopering
c. White coopering

Dry Coopering

Dry coopering centered around building casks and buckets that did not need to hold liquid. These containers were the forerunner of the modern cardboard box, transporting goods such as apples, nails, and grain. This form of coopering required the least amount of skill and the lowest quality of wood. Dry-coopered containers had thin staves and were only used once.

Wet Coopering

Wet coopering centered around building casks that were watertight. Wet coopering required the highest-quality wood and the greatest amount of skill.

The main coopers remaining today are wet coopers. Plastic and metal containers have made the wet cooper obsolete in all areas except one, and that is in the fermentation process. Wine and liquor distilleries still require oak casks for their products' fermentation and storage, as do the manufacturers of Worcestershire sauce, Tobasco® sauce, wine vinegar, etc.

White Coopering

White Coopering centers around the wooden bucket and similar items. It gets its name from the fact that these items were used to transport and store milk, sugar, and flour. It was considered the intermediate form of coopering because it required enough skill to make an item watertight, but did not generally require one to know how to bend the staves, as is required in wet coopering. It was the oldest form of cooperage and did not require premium quality wood as does wet coopering.

For further information on this, I recommend you read Kenneth Kilby's book *The Cooper and His Trade*, which covers this much more comprehensively than we can here (see Bibliography for more information).

Chapter 3
Nomenclature

The following drawings show the parts of a wooden bucket. These parts will be referred to throughout the remainder of the book, so be sure to become familiar with these terms.

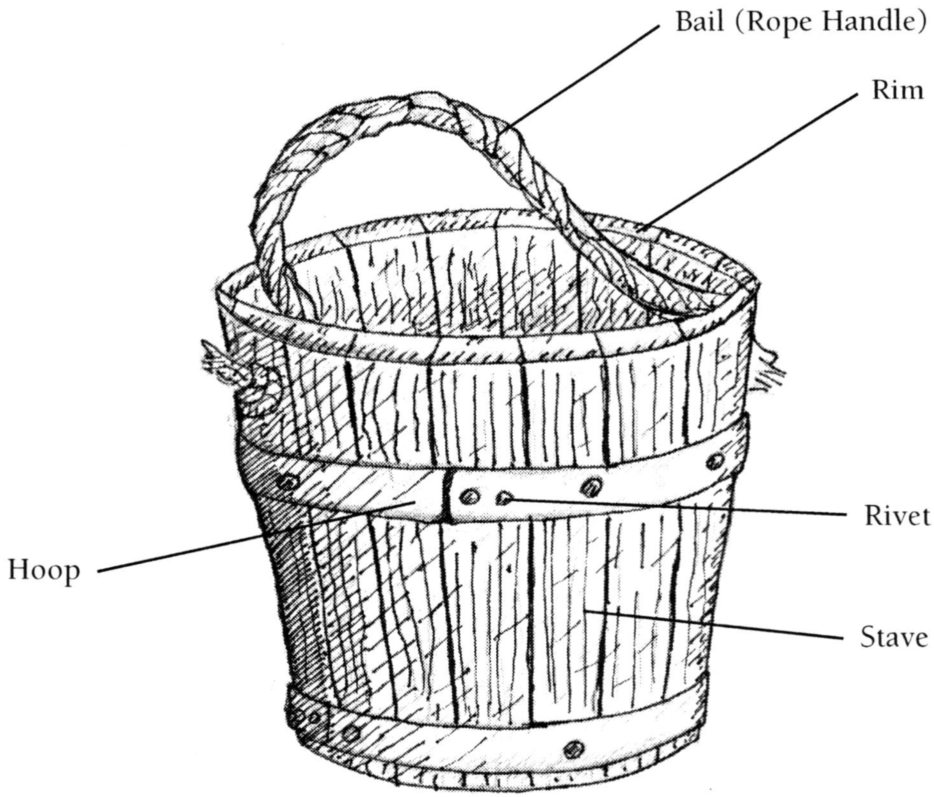

Cross Section of Bucket

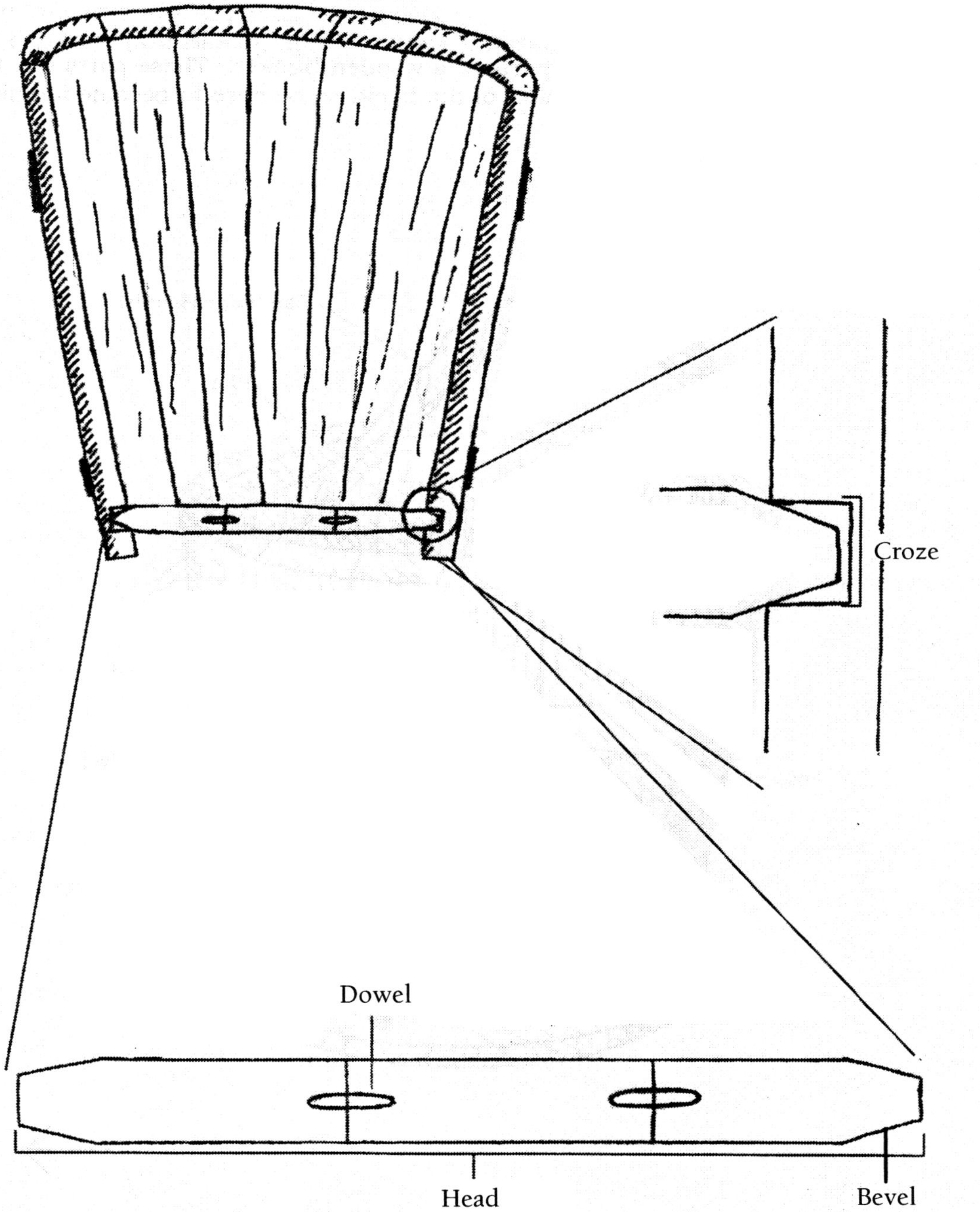

Chapter 4
Tools and Materials Needed

Tools List

The following is a list of tools that you will need to create a wooden bucket. The first list will be of tools you will be using that one can find in a modern-day woodworker's toolshop.

- Crosscut Saw
- Folding Rule or Tape Measure
- Pencil (or Marker)
- Slegehammer
- Splitting Wedge
- Wooden Mallet or Maul
- Drawknife
- Coping Saw
- Cord or Rope
- 16" Plane
- Cloth Sack or Old Pillow
- Tailor's Tape
- Anvil
- Two-pound Steel Head Hammer
- Center Punch
- 12 oz. Ball Pein Hammer
- Carpenter's Marking Gauge
- Dividers
- Small Square
- Brace
- File or Fine Wood Rasp
- Small 12" Level
- 3/8" Twist Auger
- Clean Cloth (optional)
- Double Boiler (optional)

Obsolete cooper's tools used in this book

In addition to these tools, coopers used a number of tools that are permanently obsolete, and so rare that today's hobbyists are forced to either find them at antique tool auctions or else make the tools themselves. Throughout the book, I will be discussing and illustrating some of the vintage tools coopers used, but will also show you ways to improvise using modern-day hand tools.

Listed in order of use, these tools are:

Straight Froe – looks somewhat like grass scythe, except that the blade is sharpened on the wide edge at the top. The froe will be used to section the log into staves. You can order one of these from Country Workshops (see Bibliography for more information). A substitute for the froe would be a single-bit axe or wood-splitting wedge.

Shaving Horse – The cooper's shaving horse is essentially his workbench, with one addition: it has a built-in clamp on the front for holding staves. The cooper can increase and decrease pressure on this clamp with his foot, which allows him to use both hands to dress the staves. You can also order one of these from Country Workshops (see Bibliography for more information). A substitute would be a clamp or vise and a workbench.

Hollowing Knife – Looks and works like a modern-day drawknife, except that the blade is rounded instead of straight. The hollowing knife is used to round out the concave side of a stave. Subsitutes would be a scorp or rounded-bottom plane.

Cooper's Axe – Looks like a regular hatchet, but with one very important variation: it is straight on one side so the cooper can have more control in dressing the joints of a stave (see drawing). The substitute would be a regular hatchet.

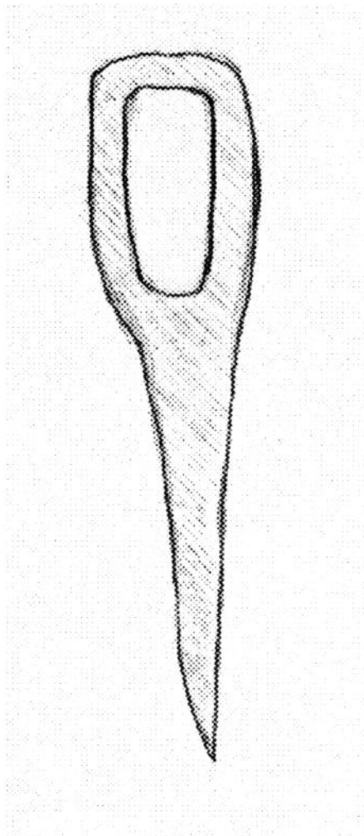

Cooper's Axe
(top view)

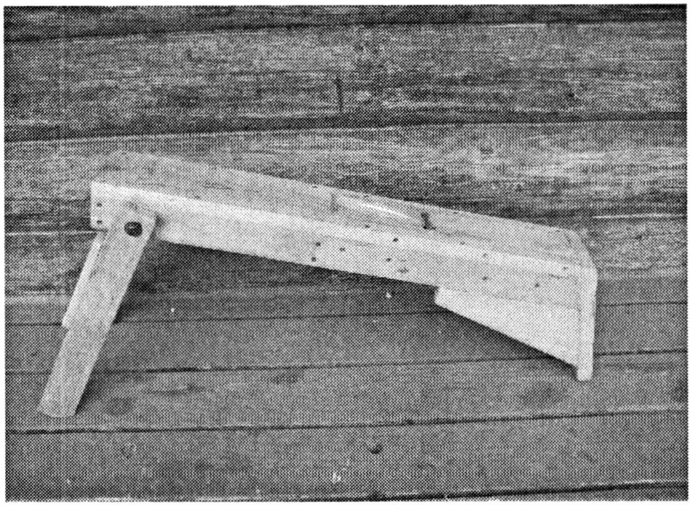

Jointer – An upturned plane about 6" to 8" long, used to plane the edge of a stave uniformly. A sharp hand plane will also work.

Hoop driver – A chisel-like instrument with a blunt, square end used for driving hoops onto the bucket. This tool can easily be made by blunting the end of a steel chisel. Or you can use a square steel shaft or a bolt head, as long as it catches the edge of the hoop.

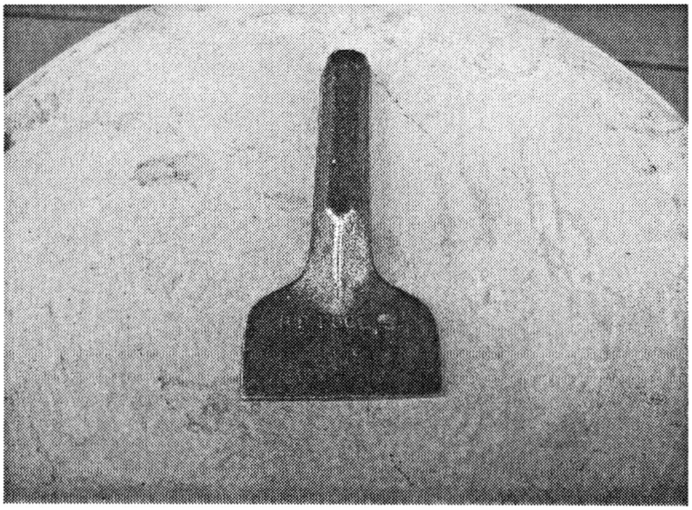

Scorp – A hoop-shaped carving knife, used to scrape out and smooth the inside of the bucket. A plane with a rounded bottom will work as a substitute.

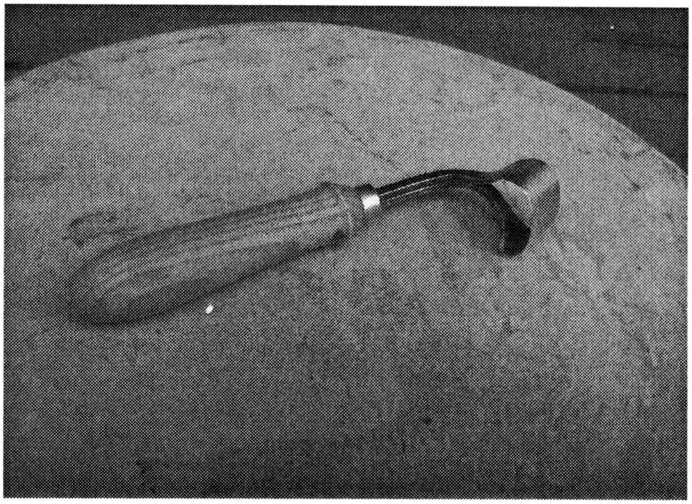

<— Side View

Top View —>

Croze – A tool that cuts the groove for the head. **Note:** I recommend that you make one following the instructions found on page 381 of the "Butter Churns" article in *Foxfire 3* (see Bibliography for more details). Otherwise, I have provided instructions for using a sharp knife or carving tools in place of a croze.

Materials

In addition, you will need the following materials:

A straight-grained log approximately 16" in diameter
16-gauge steel banding (3/4"-wide mild steel)
1/8" Rivets
1/4" hardwood dowelling
Flagging*
3/8" manila rope
Linseed oil or mineral oil (optional)
Wax (e.g. candle wax or beeswax) (optional)

*About flagging: Flagging is a cooper's term for cattail leaves or a similar plant leaf material. This material was used as a primitive sealant in the croze and in the joints of a multi-piece head (see FIGURE 6–66 in Chapter 6 for a photograph). The flagging would swell when the bucket was filled with water, thereby sealing the joints.

You will also need two *stave gauges* for measuring the angles of the staves, which will be described in the next chapter.

Chapter 5
Making Stave Gauges

Start with drawing two circles on a piece of construction paper, one 10" in diameter and one 9" in diameter. The 10" will be the diameter of the top of the bucket, and the 9" will be the diameter of the bottom of the bucket. Find the center of each circle. Draw a straight line from the center to the perimeter of the circle. This will give you the angle of the edge of the stave. Divide the circle up into twelve equal "pie slices". Now cut out one of these slices with a scissors (see Figure 5–1 for a diagram). Lay it on a piece of wood, then trace the widest part of the pie on the wood. Then cut the angle and the circle out. This will be used for measuring the convex roundness and the angle of each of the staves. The final product should look like Figure 5–2.

I have also provided patterns for the 10" and 9" stave gauges, which are on pages 28 and 29.

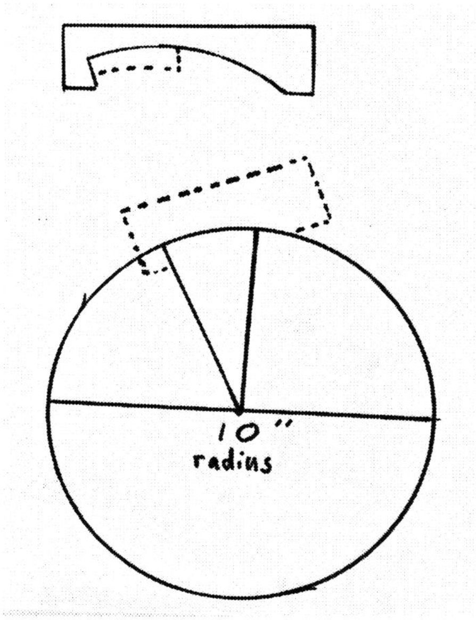

Figure 5–1

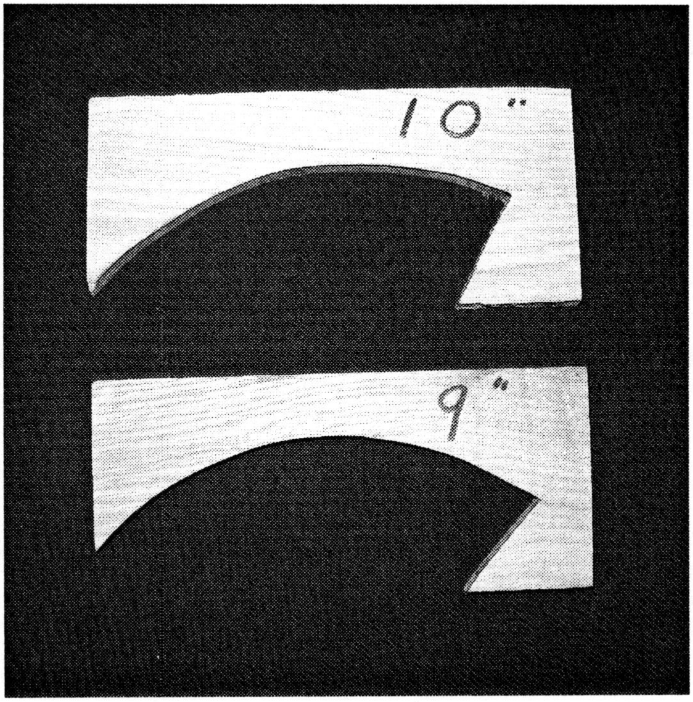

Figure 5–2

Nine-inch stave
gauge pattern

Ten-inch stave
gauge pattern

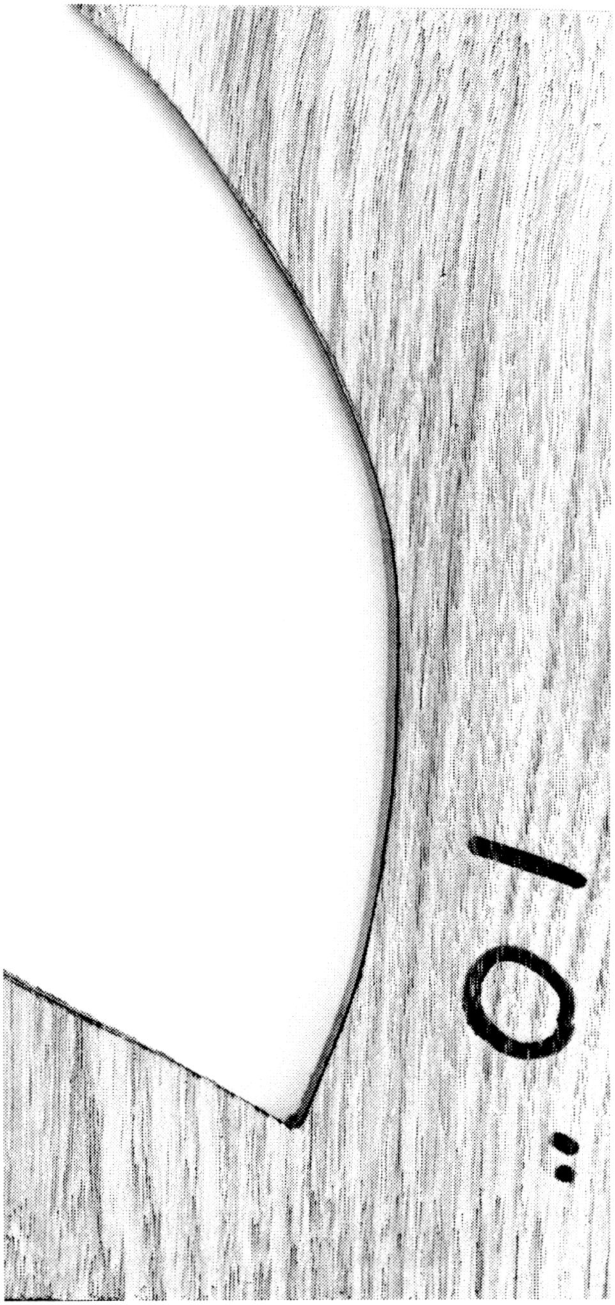

Chapter 6

Making a Wooden Bucket: The Process in Detail

In this section, we will work you entirely through the creation of a two-gallon version of The Primitive Camp Bucket, my company's best-selling item and most useful wooden bucket. This bucket has made its way into attractions such as Mt. Vernon, Lincoln Log, and the Mill Grove Audubon Wildlife Sanctuary, and has made appearances in the Imagine/Touchstone film *The Alamo* (2004).

Unlike other forms of woodworking, there are a number of variables involved in making coopered containers, so many times exact measurements are not present in the instructions; however, we have tried to include as many exact measurements as possible. Follow the instructions carefully, and with time and practice you will get a feel for the process. The completed bucket will stand approximately 10" tall, with a diameter of 10" at the top and 9" at the bottom.

(It should be explained that we have adapted this process to the modern workshop, so although the bucket itself should look authentic, much of the *process* we are using is not authentic. This is because many traditional tools and procedures are not necessarily available or practical today.)

Part One: Selecting Wood

The traditional cooper did not have the luxury of kiln-dried wood as we do today. Depending on the circumstances and area in which he was based, he might be able to have pre-split wood brought in, but many times he would have to cut the tree himself.

The cooper would select a straight-grained, easy-splitting wood relatively free of knots. White oak is the preferable wood type; however cedar and white pine are also good choices—the lighter and more decay resistant the wood the better. We will begin instruction assuming the log has already been felled.

Step 1

Cut a log approximately 16" in diameter into sections 10 1/2" long with a crosscut saw. You will want these sections to be 1/2" longer than the intended length of the bucket's staves. This extra half-inch will give you leeway in leveling the top and bottom edges of the bucket. For our Primitive Camp Bucket, the desired section length will be 10 1/2". (Figure 6–1)

Figure 6–1

Figure 6–2

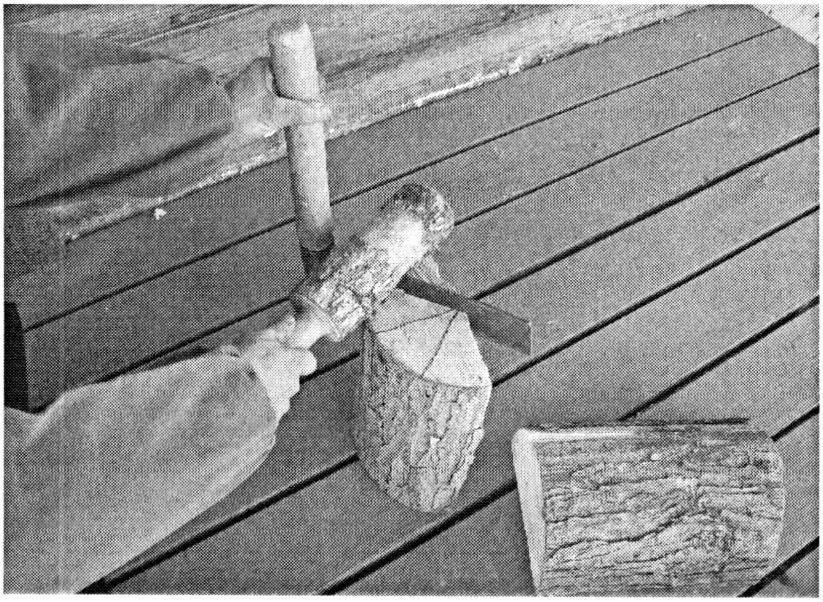

Figure 6–3

Step 2

Split the sectioned log in half using a sledgehammer and a splitting wedge. (Figure 6–2)

Step 3

Here you will begin splitting staves from the halved sections with a straight froe. Position the froe on the halved section so that it will cut off a 1 1/2"-thick wedge. (The wedge always needs to be a little thicker than the intended final thickness of the stave.) You will hold the froe by the handle with one hand (the froe's handle will be to the outside of the section). With the other hand, you will hammer the froe's blade into the wood with a wooden mallet or maul. Repeat this process until both of the halved sections have been broken entirely into staves. (Figure 6–3) If you are only splitting enough staves for one bucket, you need to have at least fifteen (15) 3 1/4"-wide staves (a few of these will be extras in case of mistakes or defective staves). If you do not have a froe, you could also split the sections with a short-handled sledgehammer, using an old axe head as a wedge.

You will need to cut the wood in the windmill fan pattern shown in Figure 6–4. This type of cut is called a "quarter cut". It will keep the wood from warping and twisting.

Step 4

Discard the outside sapwood and the inside heartwood on each stave (see stave diagram in Figure 6–4). The best wood is the wood in between these two, as this is the strongest part of the tree and the most likely to be free of knots. Cut off the rest of the excess wood with a hatchet or froe, making each section into a 1 1/8" x 3 1/4" x 10 1/2" stave (approximately). The extra width and thickness will allow for shrinkage as it dries. Then shave each board smooth using a drawknife.

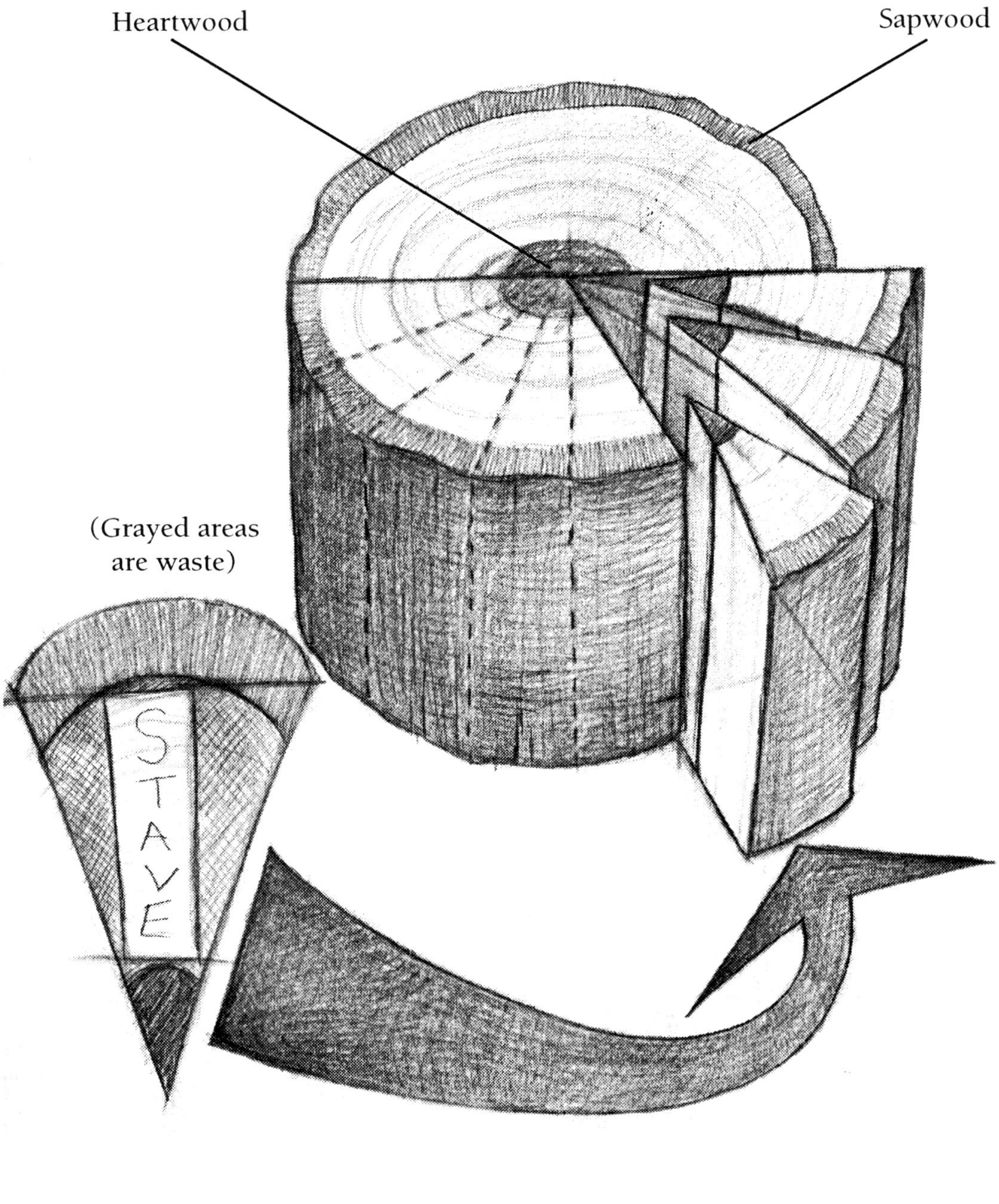

Figure 6–4

Step 5

After splitting the wood, allow the wood to dry for at least six months if you live in a dry area, or one year if you live in a humid area. The wood should be stacked in a dry, shady place with plenty of air circulation, such as inside a shed. When stacking the wood, lay small slats of wood about 3/4" thick between each layer of wood in the pattern shown in Figure 6–5 to allow for proper air circulation.

Figure 6–5

Part Two: From Staves to Bucket

A. Shaping and Binding the Staves

Step 6

Once the staves have dried, shave the face and back of each stave into a curved shape. The cooper "hollows" the face (or concave side) with a hollowing knife (Figure 6–6), and shaves the back (or convex side) with a straight drawknife Figure 6–7). If you do not have a hollowing knife, a scorp or a rounded hand plane will work as substitutes.

The cooper accomplishes this step by sitting at the cooper's shaving horse with the stave held by one end in the clamp. A reasonable substitute would be to clamp the jaws of a bench vise or a C-clamp onto a sawhorse or bench (although you will want to pad the jaws of the clamp with some type of softwood). **NOTE**: Whether using a shaving horse or a clamp, be careful not to clamp the stave too tightly or else it will break.

Figure 6–6 (above); Figure 6–7 (below)

Figure 6–8

Step 7

Rough-cut the joints of each stave with a cooper's axe, giving each a straight inward taper from top to bottom so the stave will have a slight wedge-like appearance. (Figure 6–8) The more of a difference in diameter there is between the top and bottom of the bucket, the more wedge-shaped the taper will need to be. For our purposes, the stave needs to be 3" wide on one end and 2 1/2" wide on the other end. (Figure 6–9)

Figure 6–9

Step 8

You will need to assemble a form to build the bucket around. Measure out and cut two circles from 3/4"-thick pine or 3/4" plywood. One circle should be 7 1/2" in diameter, and one 8 1/2" in diameter. Then cut a section of a 2 x 4 or 4 x 4 board 8 1/2" long; this will be a spacer. (Figure 6–10) Nail the two circles to the opposite ends of the spacer, making sure that the center of each end of the spacer is nailed to the exact center of each wood circle. The final form should look like Figure 6–11. You could also use an existing bucket, such as a metal pail 7 1/2" on top and 8 1/2" on bottom, as a substitute.

NOTE: Whether using a wooden form or another bucket, make sure the small end is pointing up so that gravity will be on your side when you lay the staves on the form.

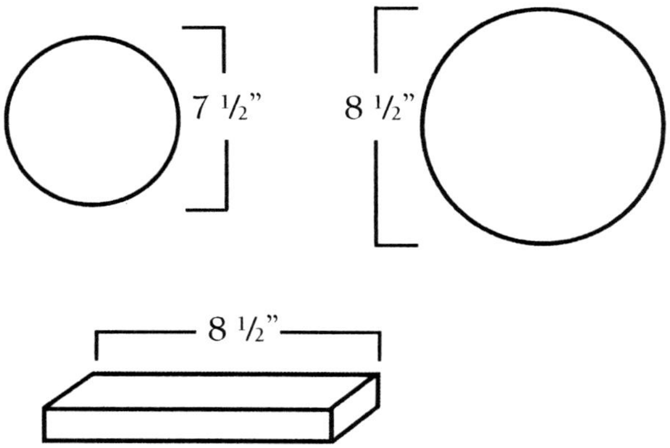

Figure 6–10

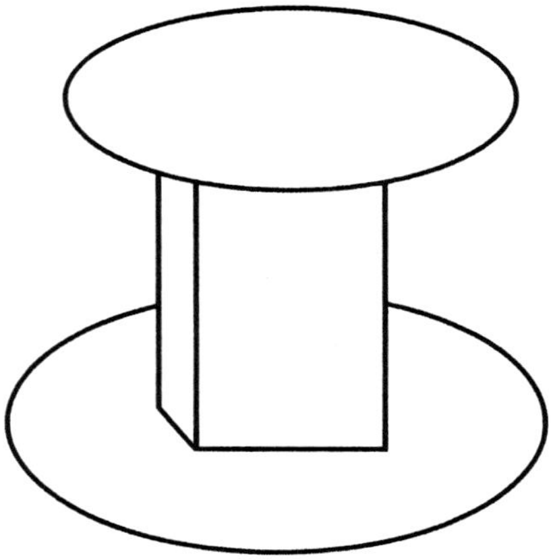

Figure 6–11

Step 9

Rough bevel the edge joints of the first stave with a jointer. (Figure 6–12) A substitute for the jointer would be a plane turned upside down and clamped securely in a bench vise. When you are jointing the stave, make sure the top of the stave is pointed away from you. Then start increasing the pressure toward the concave side after you get past the first third of the stave and follow through with this extra pressure to the end of the stave, thereby blending the two angles together. (**NOTE:** this is a difficult but important step, and may require a little practice on a scrap board.) Repeat this process on the other edge. Each side of the stave should have a bevel of 9°, with the angles tipping inward. (Figure 6–13) Lay the finished stave against the form. The narrow end of the stave will be the narrow end of the bucket, and the wide end will be the wide, open end of the bucket.

Figure 6–12

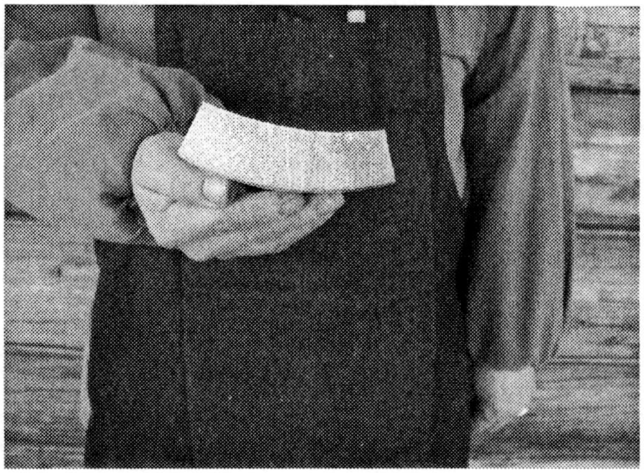

Figure 6–13

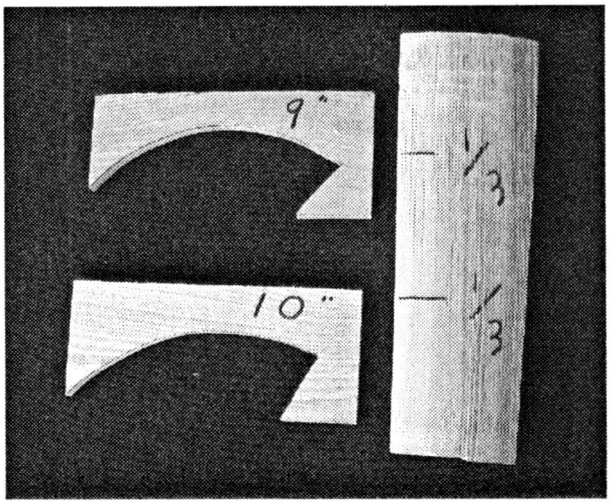

Figure 6–14

As you joint the stave, check the stave's angles with the stave guides (Figure 6–14). Check the top one-third of the stave with the 10" guide (Figure 6–15) and the bottom one-third with the 9" guide (Figure 6–16). The middle third should blend the top and bottom angles together.

CAUTION: Keep your fingers as far away from the jointer blade as possible.

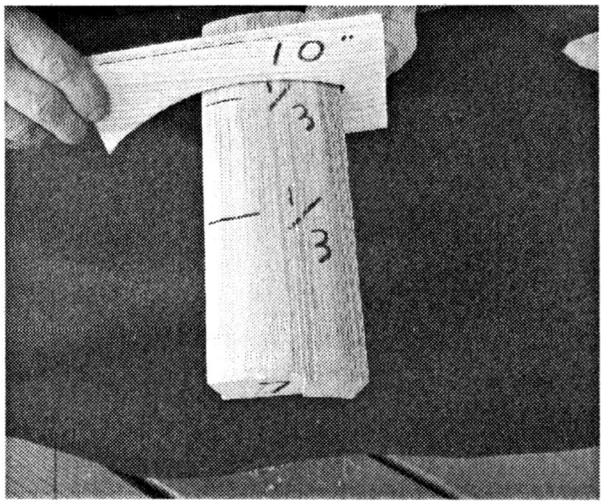

Figure 6–15

Figure 6–16

Figure 6–17

Figure 6–18

Step 10

Repeat Step 9 with the next stave. Lay this stave next to the other on the form. As you do this, check the joints for gaps. Each stave should lay smoothly and evenly against the next stave. If there are any gaps between staves, adjust the joints as necessary with the jointer or plane. In order for the bucket to hold water, you must lay each stave tightly and smoothly against the next stave all the way around the form. Any extra staves you made will come in handy here if any stave is too narrow or has a defect. This is probably the most time-consuming step of the process, but it is necessary in order for the bucket to be watertight. Continue this process, working your way around the form (See Figure 6–17) until only one space is left.

Hint: Number the staves as you go along, so if the staves come apart, you will be able to reassemble them around the form in the right order. (Figure 6–18)

NOTE: The staves must be bound together while you are placing them around the form to keep them in place. We've found that a cord with a slipknot tied in the end to make a noose works the best. As you add the staves, you can loosen or tighten this noose as needed.

Step 11

The last stave will have to be specially cut to fit the final remaining space on the form. First, measure the remaining space (Figure 6–19). Then apply the measurement to the final stave (Figure 6–20) and bevel the last stave accordingly. It is better to cut this stave a little too wide than a little too narrow. If it is too wide, you can plane it to fit. Then slide this final stave into the space (Figure 6–21).

Figure 6–19

Figure 6–20

Figure 6–21

Step 12

Carefully flip the bucket upside-down (Figure 6–22). Then lift the form out of the bucket, taking care to keep the rope around the staves tight (Figure 6–23). Then insert a cloth sack or old pillow inside the bucket in place of the form (Figure 6–24). This soft replacement "form" will hold the staves together from the inside.

Figure 6–22

Figure 6–23

Figure 6–24

B. Making and adding temporary hoops

When making a bucket, you will need three temporary hoops. One temporary hoop will need to be large enough to go three quarters of the way down from the narrow end of the bucket. The next temporary hoop will be smaller, fitting approximately one quarter of the way down from the narrow end of the bucket. The final temporary hoop will go around the exact middle of the bucket. The temporary hoops will end up having a slightly larger circumference than the permanent hoops (which you will be making in Steps 32–36) because you will be shaving down the outside of the bucket before making the permanent hoops, which will shorten the circumference of the bucket.

Here I will show you how to make the three temporary hoops. These hoops can be used again on later projects.

Step 13

Measure the bucket from top to bottom. Find the quarter, half, and three-quarter marks on this measurement and mark each of these increments on the bucket. (Figure 6–25)

Step 14

Measure the circumference of the bucket at the halfway mark with a tailor's tape–this will be measurement number 3. (Figure 6–26) Then measure the circumference of the bucket at the one-quarter measurement (this will be measurement number 1) and at the three-quarter measurement (this will be measurement number 2). Record all these measurements.

Figure 6–25

Figure 6–26

Step 15

Here you will be cutting three lengths of steel banding, one for each circumference measurement you recorded in Step 14. Use 3/4"-wide, mild 16-gauge steel banding or a similar material. First clip the end of the banding so that it is square. Then transfer measurement #3 onto the banding and make a mark on the banding at the end of the measurement. (Figure 6–27) Make another mark 1 1/2" beyond this measurement. (Figure 6–28) Mark the middle hoop with the number "3" (Figure 6–29), as it will be referred to as Temporary Hoop #3 from now on. Then cut the banding at the second mark (Figure 6–30)–this will give you the circumference of the middle hoop plus a 1 1/2" overlap. Repeat this process with measurements #1 and #2, making sure each cut is square. The small one-quarter measurement hoop will be Temporary Hoop #1, and the large three-quarter measurement hoop will be Temporary Hoop #2; be sure to mark each accordingly.

Figure 6–27

Figure 6–28

Figure 6–29

Step 16

Here you will need an anvil or a metalworking surface of some kind. Lay temporary hoop #3 over the anvil and hammer out the metal on one edge (Figure 6–31). This process is called "bruising" or "stretching" the metal; it will give the band a half-moon or "smile" appearance (Figure 6–32). This will enable the metal to wrap naturally with the shape of the bucket. Repeat this process with Temporary Hoops #1 and #2.

Note: You don't have to heat the metal to do this; the metal can be successfully bruised when it is cold.

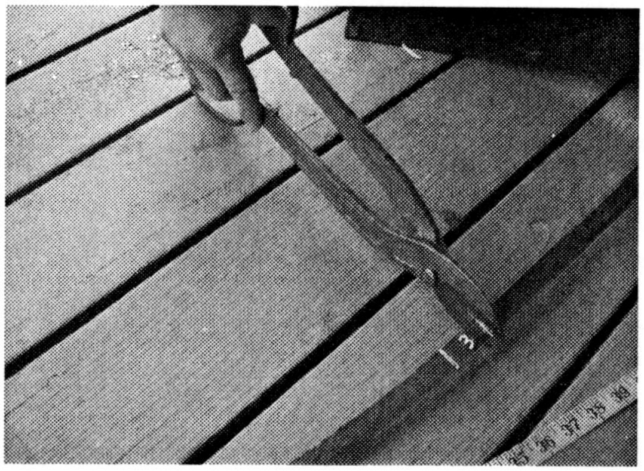

Figure 6–30

Figure 6–31

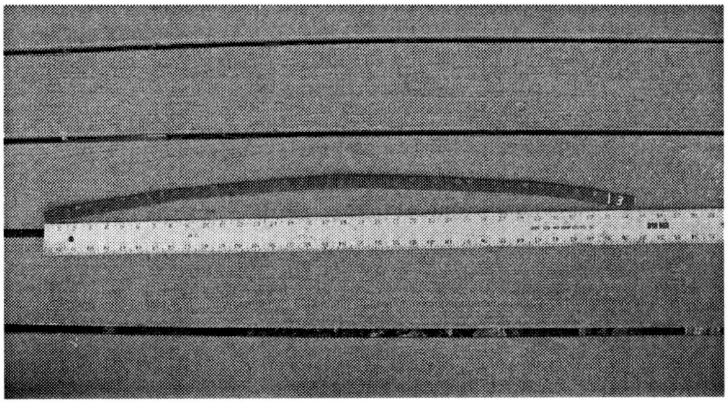

Figure 6–32

Step 17

Bring the two ends of Temporary Hoop #3 together to make a circle. (Figure 6–33) Bring the one end to rest at the first marking that you made in Step 15 so that the metal will overlap 1 1/2". (Figure 6–34) Then rivet the ends together *once* in the middle of the overlap, first punching a hole into the metal as shown in Figure 6–35 with a center punch and a two-pound steel hammer, and then hammering in the rivet as shown in Figure 6–36 with a ball pein hammer. The reason you will only rivet the hoop once is so that the rivet can act as a hinge, allowing for a more flexible hoop. Repeat this process with Temporary Hoops #1 and #2, making sure each overlaps 1 1/2".

NOTE: After hammering the rivet tight, smooth the inside of the hoop by hammering the rivet and joint against the anvil so that there will be no rough edges that could potentially scar the wood.

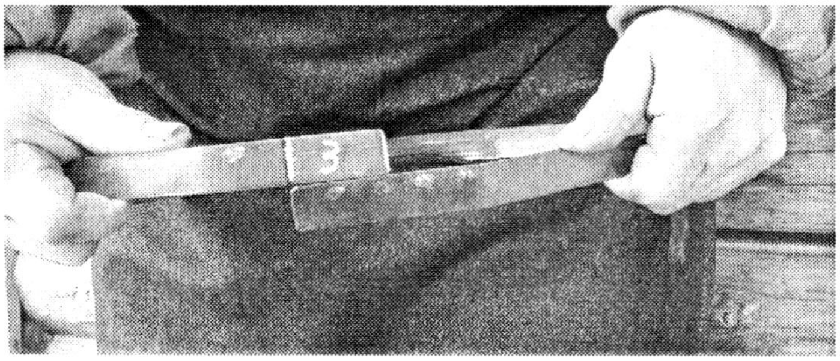

Figure 6–33

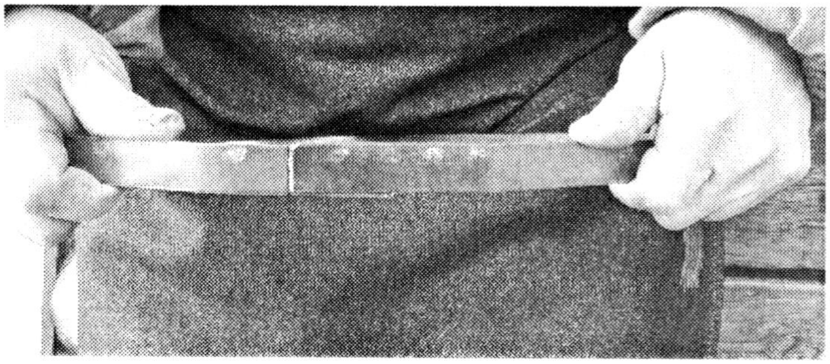

Figure 6–34

Figure 6–35

Figure 6–36

Step 18

Lay Temporary Hoop #3 aside; you will not be using it until Step 31. Drive Temporary Hoops #1 and #2 over the staves with a two-pound steel head hammer and hoop driver until they fit snugly over the bucket. (Figure 6–37) If you do not have a hoop driver, a bolt head or a square steel shaft will do, as long as it catches the edge of the hoop without doing much damage to the wood. The bruised edge of the hoop should point toward the wider end of the bucket as pictured in Figure 6–38. As you are driving the hoops down, you should alternate driving positions on the hoop to keep the hoops even all the way around.

NOTE: Do this on a flat surface and make sure the bucket is straight and not lopsided (some of the old coopers called a lopsided bucket a "drunken" bucket, others called such a bucket a "lord" from "drunk as a lord"). (Figure 6–39)

Figure 6–37

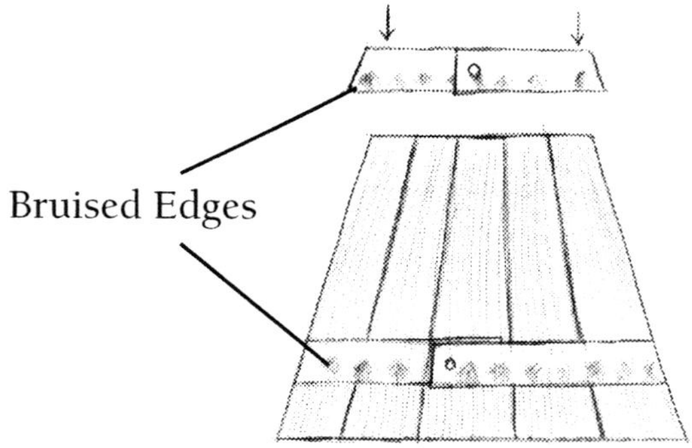

Figure 6–38

Lopsided or “drunken” bucket

Straight bucket

Figure 6–39

C. Cutting the croze

Step 19

Remove the soft "form" from inside the bucket. (Figure 6–40) Smooth the inside of the bucket with a scorp until it is completely rounded and smooth on the inside. (Figure 6–41)

NOTE: It is important that the inside of the bucket be as round as possible; this way the head will fit properly in the croze and the head will not leak. This can be done by taking the 7 1/2" wooden circle off of the form you made in Step 8 and using it as a circular gauge. Insert the 7 1/2" circle into the wide end of the bucket, and slide it down toward the narrow end. Shave off all the high spots that will keep the circle from slipping through the narrow end of the bucket (See Figure 6–42 for a visual guide).

Figure 6–40

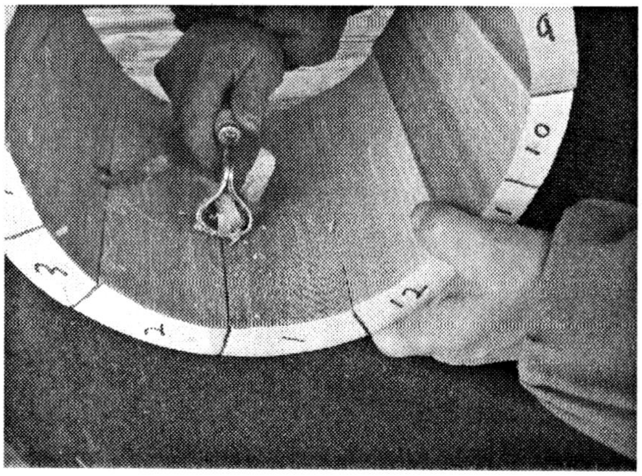

Figure 6–41

Before

After

Figure 6–42

Step 20

Level the bottom of the bucket with a plane so that all the staves are even and level. (Figure 6–43) Old coopers would use a topping plane for this, but an ordinary steel plane will work just as well.

NOTE: Make sure the bottom of the bucket is completely level and smooth.

Step 21

At this point, you will need to have a croze that will cut a square notch (as compared to a v-notch), or a sharp pocket knife or carving tool set. Step 21 will be written for both of these applications. Either way, you will end up with a croze that will hold the head 3/4" above the ground, which will keep the stress on the staves and not on the head.

With croze:

Lay the croze squarely on the leveled edge of the bottom of the bucket and swiftly run the croze around this edge, gently pushing inward and downward. (Figure 6–44) As with a saw, you do not want to force the blades but rather allow the croze to do most of the work. This groove should be 3/4" from the bottom of the bucket so that the wood between the groove and the bottom will be strong enough to hold the head in. The groove should not be more than one-third of the thickness of the staves, otherwise it will weaken the staves too much.

Figure 6–43

Figure 6–44

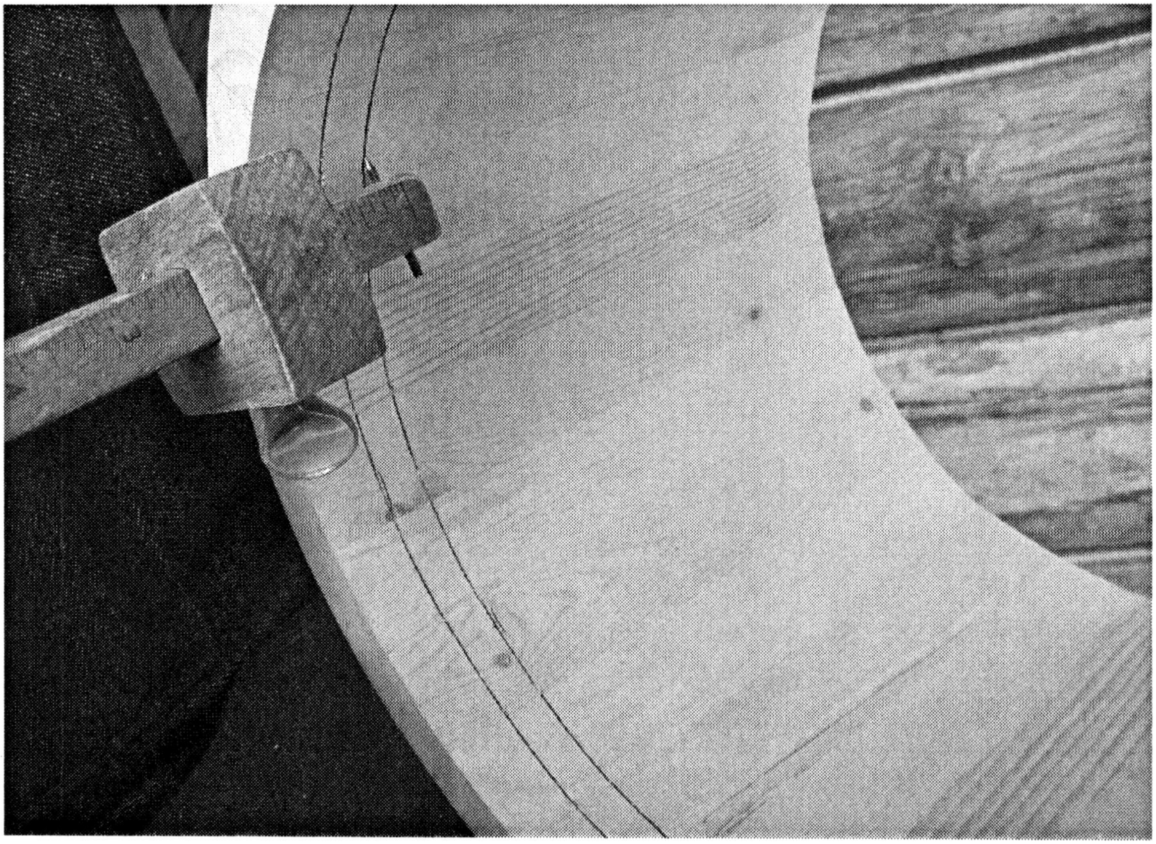

Figure 6–45

With sharp knife or carving tools:

Mark 3/4" down on the inside of the bucket with a carpenter's marking gauge. Then make another line 1 1/4" down with the carpenter's marking gauge. (FIGURE 6–45) Use a sharp knife or carving tool to cut a 1/4"-deep groove between the two lines. The groove should not be more than one-third of the thickness of the staves, otherwise it will weaken the staves too much.

D. Assembling and inserting the head

When making buckets, you can choose between making a one-piece or a multiple-piece head. A one-piece head is the easiest to make, and the uniformity of a one-piece head adds a nice touch to the bucket. However, a one-piece head will also be more susceptible to warping and twisting, whereas a multiple-piece head will be stronger and less likely to warp. On our two-gallon bucket, two boards will make a sufficient head, but on a larger bucket, three to five pieces will work better. The instructions here provide for both head types.

STEP 22

Here you will need a pair of dividers. Adjust the dividers until they will go six times around the outside diameter of the groove you made in Step 21 minus 1/8" (the 1/8" leeway will allow for room to put the flagging in the joints). This is done by sight and through trial and error. (FIGURE 6–46)
NOTE: As you get better at making buckets, you can leave the flagging and the 1/8" gap out. The flagging will decompose faster than the bucket will.

One,

Two,

Three,

Four,

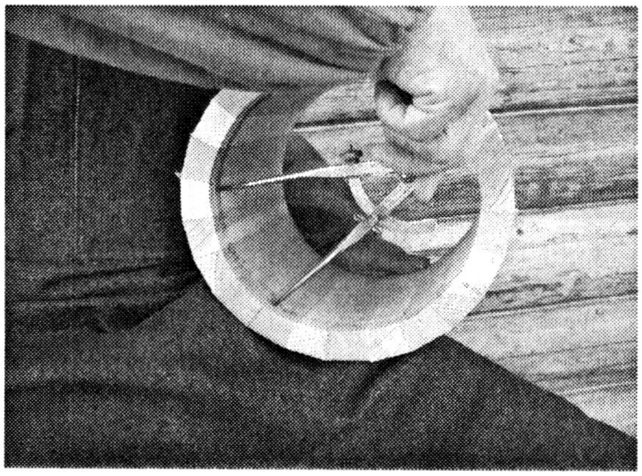

Five,

Six times—,
the sixth measurement should be 1/8" short of the full circumference.

Figure 6–46

Figure 6–47

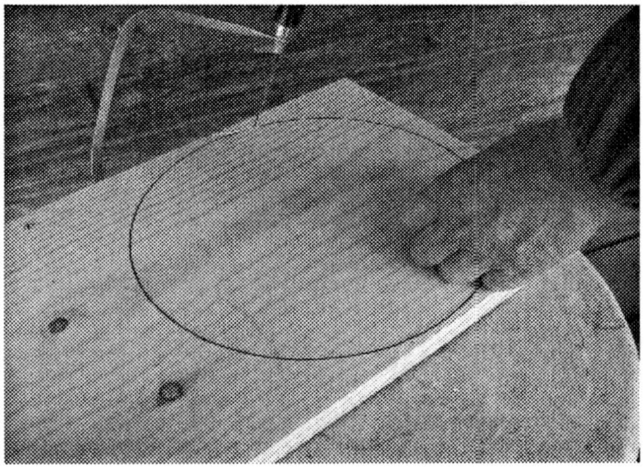

Figure 6–48

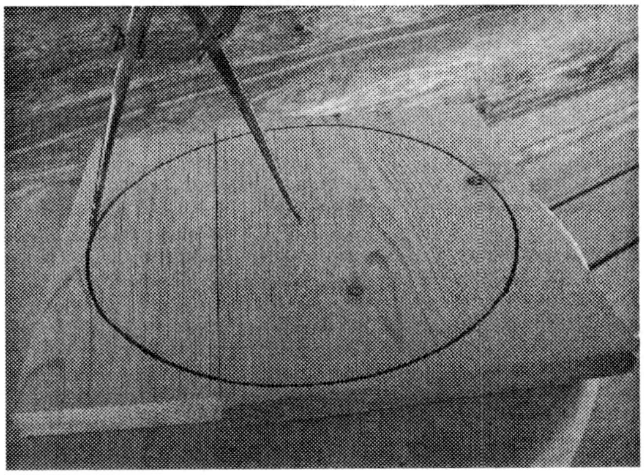

Figure 6–49

Step 23

If you are making a one-piece *head:*

Choose an 11 1/2" section of 1 x 12 lumber with no knots or warp. Draw a complete circle on the board using the measurement on your dividers from Step 22. (Figure 6–47) Cut the circle out of the board with a coping saw. (Figure 6–48)

If you are making a two-piece *head:*

a. Lay two 3/4" boards cut to 10" lengths next to each other. The exact width of the boards does not matter, as long as they add up to at least 10". (I used a 6" board and a 4" board for the pictures.) Make sure that the boards fit together squarely and are completely straight, with no warps or twists (joint each board square if necessary).
NOTE: You can make these boards out of any remaining unshaped staves.

b. Hold the boards together firmly and draw out a circle on the boards using the measurement on the dividers from Step 22. (Figure 6–49)

c. The head boards will be held together by a series of alternately-positioned dowels. Draw lines over the joints to position the dowels following the pattern shown in Figure 6–50.

d. Transfer the dowel lines over the edge of each board (Figure 6–51). (I use a small square to accomplish this.) These lines should be drawn across the entire thickness of the board. Then make a mark in the exact middle of each line (Figure 6–52). This will be where the dowel holes will be drilled.

e. Drill the dowel holes (Figure 6–53). The old coopers used a spade bit and a wood brace, although I have chosen to use a modern drill bit because it works better. The alternative to a wood brace is to use a carpenter's brace with a 1/4" drill bit; hold the boards with a wooden clamp if necessary. Drill each hole 3/4" deep.

Figure 6–50

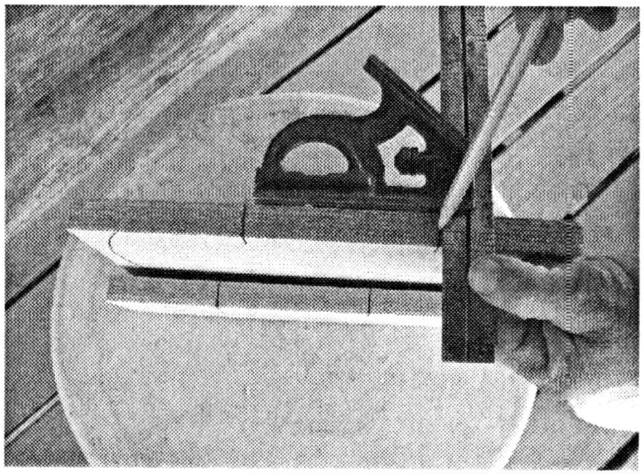

Figure 6–51

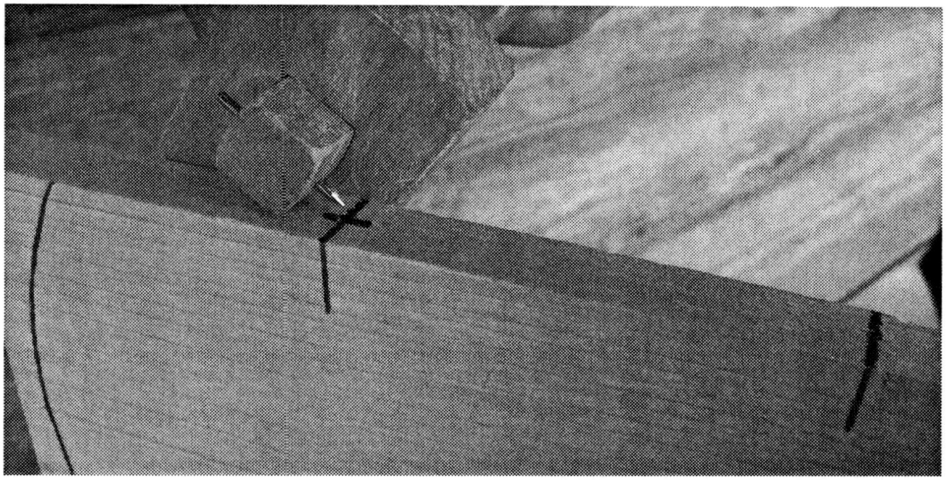

Figure 6–52

Figure 6–53

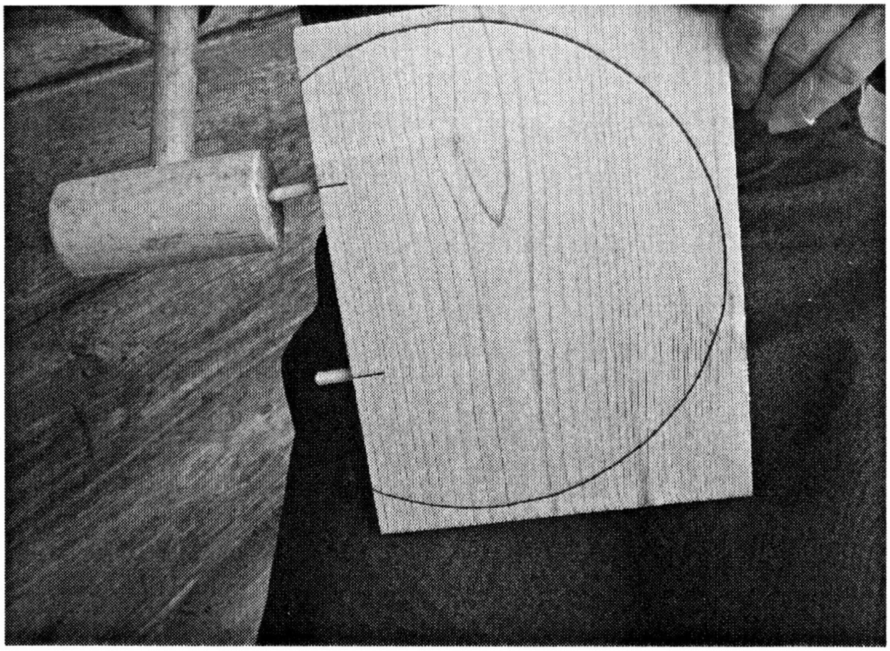

Figure 6–54

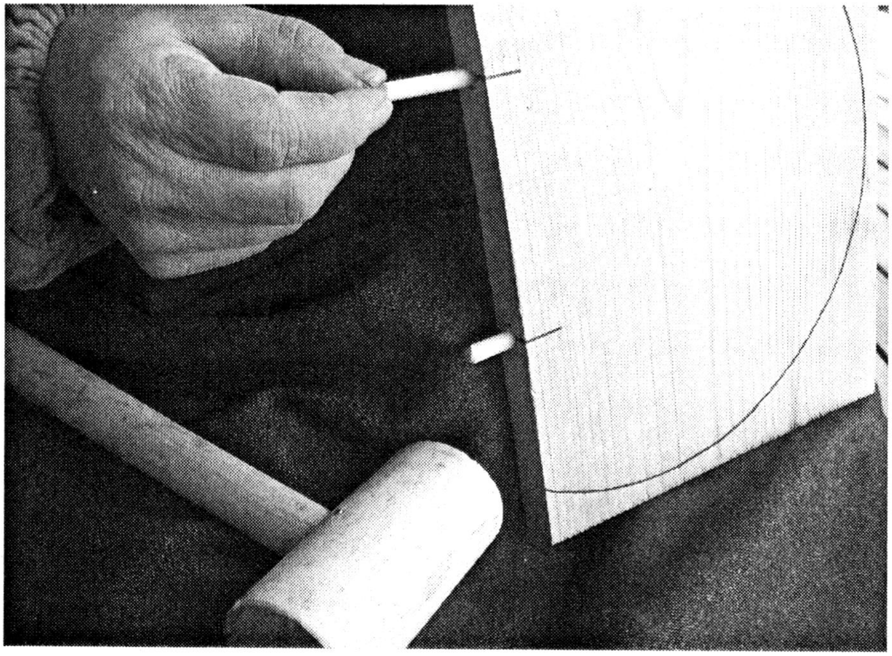

Figure 6–55

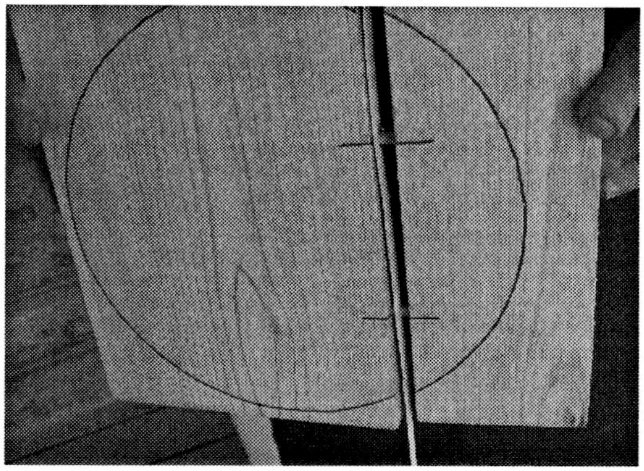

Figure 6–56

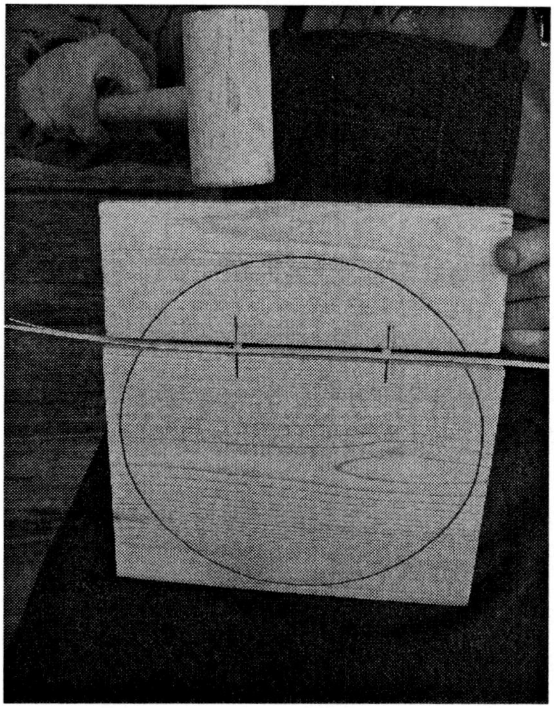

Figure 6–57

f. Cut two 1 1/2" dowels from 1/4" hardwood dowelling. Insert the dowels into one of the drilled boards. (FIGURE 6–54)

g. Drive the dowels into one of the boards with a wooden mallet. (FIGURE 6–55) Lay reed flagging (see page 25 for more information on flagging) on the corresponding edge of the other board. Position the two boards together (FIGURE 6–56), and then drive them tightly together with a wooden mallet. (FIGURE 6–57) The dowels will, of course, break through the flagging, but be sure the dowels do not break the flagging completely in two. The flagging will work as a sealant.

h. After reassembling the boards, redraw the circle using the divider measurement from Step 22 if the circle is no longer true (FIGURE 6–58), then cut out the circle with a coping saw (see FIGURE 6–48 on page 61).

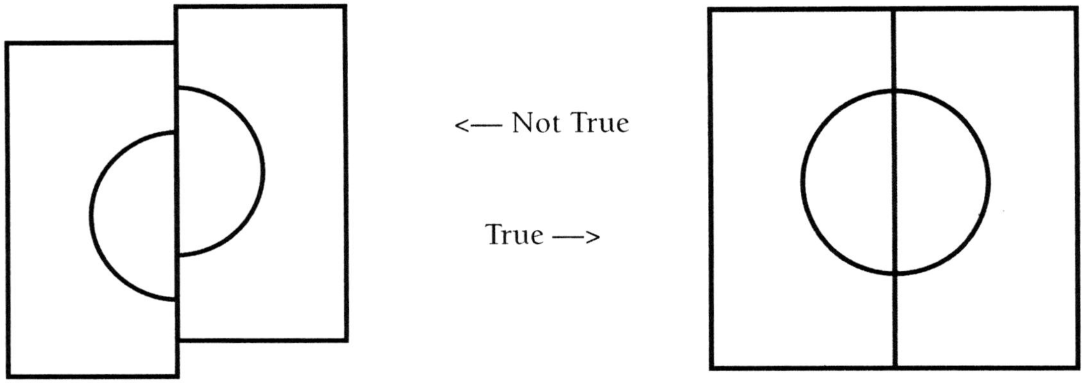

Figure 6–58

Step 24

Smooth the edge of the head with a file. (Figure 6–59) Draw a circle on one side of the head 1 1/2" smaller than the actual head (Figure 6–60), then turn the head over and draw another circle 1 1/2" smaller than the head. (Figure 6–61) These markings are where you will begin beveling the head. Now find the middle of the edge of the head, which should be at 3/8". (Figure 6–62) Draw two lines completely around the edge of the head, one 1/4" to the left and 1/4" to the right of the middle. (Figure 6–63) These lines are the boundaries the bevel will slope to.

Figure 6–59

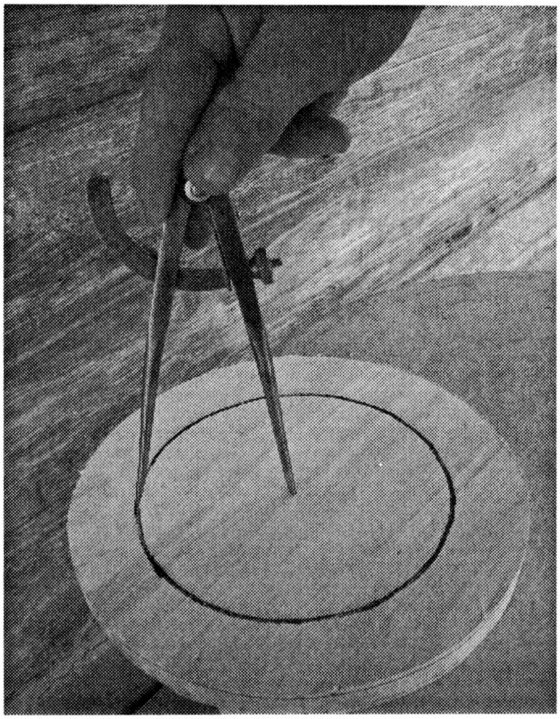

Figure 6–60

Figure 6–61

Figure 6–62

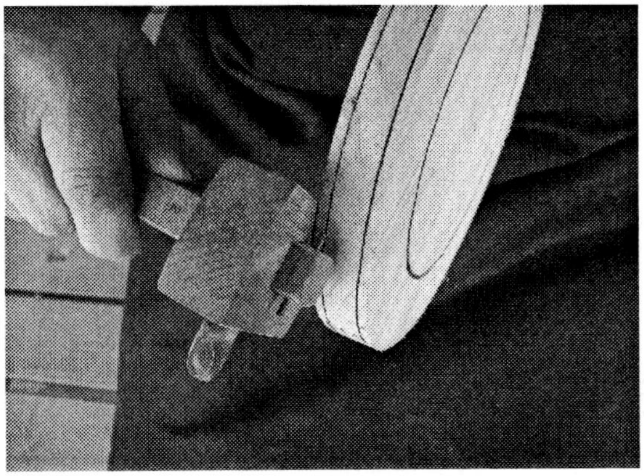

Figure 6–63

Figure 6–64

Step 25

Using a straight drawknife, bevel the edge of the head on both sides within the boundaries you made on Step 24. (Figure 6–64) Then using a file or a fine wood rasp, smooth the bevel of the head on both sides all the way around. (Figure 6–65)

Figure 6–65

Step 26

Put flagging all the way around the inside of the croze. As with the joints inside the multi-piece head, this will act as a sealant. (Figure 6–66)

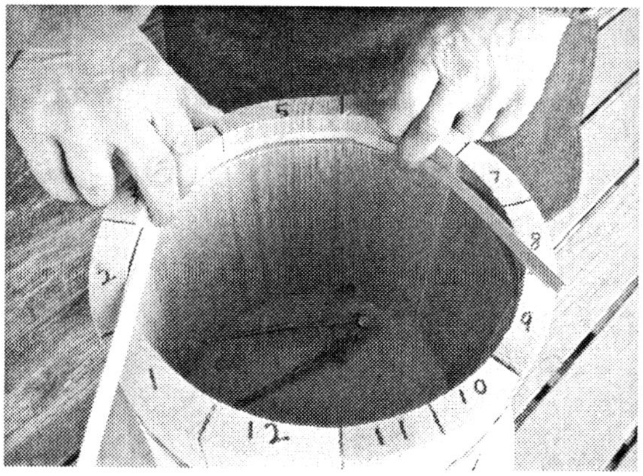

Figure 6–66

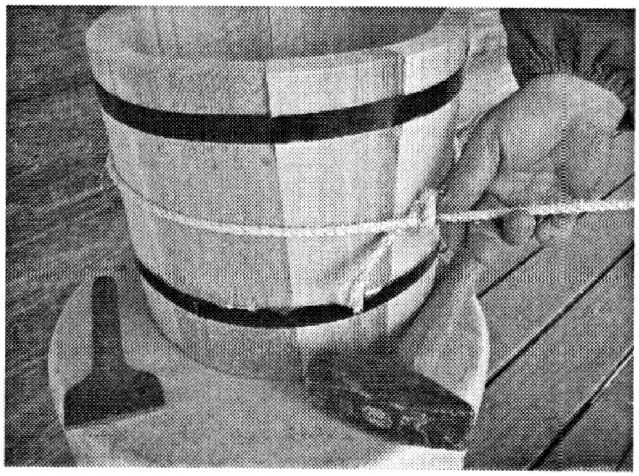

Figure 6–67

Step 27

Tie a cord around the middle of the bucket (Figure 6–67) and flip the bucket so that it stands narrow-side down. Drive both temporary hoops off the bucket. Then slip the head into the croze. (Figure 6–68)

Step 28

Flip the bucket back over. Drive both temporary hoops on as tightly as you can. (Figure 6–69) Remove the cord.

Figure 6–68

Figure 6–69

Step 29

Check the head to make sure it fits well and evenly inside the croze.

If the head is too tight (i.e. the staves will not go back together and there are new gaps between them):

a. Re-insert the soft form, tie the cord around the bucket again, loosen the hoops just enough to where they are finger-tight, and lift the head out of the bucket.

b. Rasp the edge of the head down slightly all the way around.

c. Carefully slip the head back into the bucket. Drive the hoops back on until they are tight. Remove the soft form and the cord.

If the head is too loose (i.e. it slides around inside the croze):

a. Re-insert the soft form, tie the cord around the bucket again, and then carefully loosen the hoops until you can pull out one stave.

b. Shave the joint of this stave slightly with the jointer or a hand plane to make it narrower. Slip it back in with the other staves.

c. Drive the hoops back into place. Remove the soft form and the cord.

After you have driven the hoops back into place, check the fit of the head again. Repeat this process again until the head fits correctly.
NOTE: As time consuming as it is, this is a CRITICAL STEP if the bucket is to hold water.

Figure 6–70

E. Smoothing the outside of the bucket

Step 30

Smooth the outer midsection of the bucket between the top and bottom hoops with a fine wood rasp. (Figure 6–70)

Step 31

Drive Temporary Hoop #3 over the middle of the bucket. (Figure 6–71) It should be snug, but not so much that it will scar the wood. Drive off Temporary Hoops #1 and #2 (Figure 6–72) and, using a file or rasp, smooth the area of the bucket where Temporary Hoops #1 and #2 were. (Figure 6–73)

Figure 6–71

Figure 6–72

Figure 6–73

F. Making and adding permanent hoops

Step 32

Follow directions in Step 13 (in the "Making Temporary Hoops" section) for finding the quarter lengths for the new permanent hoops. Measure out the new circumference of the bucket at both quarter measurements, adding a 1 1/2" overlap to each measurement. These measurements must be exact. **NOTE:** You will only be making *two* hoops, one at the quarter measurement and one at the three-quarter measurement. You will not need to make a third hoop at the half-measurement.

Step 33

Repeat Steps 15 and 16 for making the upper and lower permanent hoops.

Step 34

As in Step 17, bring the two ends of steel together to form a circle, making sure that the steel overlaps 1 1/2". Then rivet each hoop together. This time you will use *two* rivets instead of one. (Figure 6–74) This will make for a stronger hoop. Bend the top end of the hoop over the bottom end by hammering the two ends together on a flat piece of steel. This is so that the two ends of the hoop will fit together as smoothly as possible and will not scar the wood as you drive the hoop down (the final result should look like Figure 6–75).

Figure 6–74

Figure 6–75

Step 35

Drive both hoops on as tightly as possible. (Figure 6–76) When driven on tightly enough, the hammer blows will ring like a wooden bell.

Step 36

Flip the bucket over and remove Temporary Hoop #3 (Temporary Hoop #3 holds the bucket in place while you are driving on the permanent hoops). (Figure 6–77)

Figure 6–76

Figure 6–77

G. Smoothing the edges of the bucket

Step 37

Level the top of the bucket with a plane. (Figure 6–78) Again, a topping plane works best for this, but an ordinary steel plane will work also.

Step 38

Smooth the outside and inside edges of the bucket with a fine, round-faced wood rasp. (Figure 6–79) There should be no sharp or splintering edges on the bucket whatsoever.

H. Making and adding the Bail (Rope Handle)

Step 39

Measure down 1 1/2" on the exact opposite sides of the bucket with a carpenter's marking gauge. (Figure 6–80)

Figure 6–78

Figure 6–79

Figure 6–80

Figure 6–81

Step 40

Take two pointed objects (e.g. nails) and poke them into the opposite sides of the bucket at the markings made with the carpenter's marking gauge. Lift the bucket with the nails and adjust the nails horizontally until the bucket balances level. (Figure 6–81) This will give you the markings for drilling the holes for the bail.

NOTE: It will help to lay a small level on the top of the bucket to check the balance while doing this, as shown in the picture.

Step 41

Drill a hole 3/8" in diameter at each point marking you made in Step 40 with a twist auger. Drill from both sides of the stave so that you don't splinter the wood out. (Figure 6–82)

Figure 6–82

Step 42

Cut a length of 3/8" manila rope 36" long. Run the rope through the holes as shown in Figure 6–83 and tie a square knot on the outside of each hole. Keeping the rope on the inside creates less tension on the wood above the hole.

I. Finishing

Finishing (i.e. oiling and waxing) the bucket slows the shrinking and expanding of the bucket, which preserves and prolongs the life and usability of the bucket. Waxing helps prevent leakage, and oiling keeps the wood from discoloring and drying out. However, some prefer the "natural" look because it is more historically accurate. I will show you how to do both here.

Figure 6–83

Step 43

To seal and preserve the bucket:

Oiling. Rub linseed oil or mineral oil into the outside of the bucket with a clean cloth. (Figure 6–84) Wipe off any excess oil and allow the bucket to dry. (If you prefer, you can paint the bucket instead of oiling it.)

Waxing. Heat wax in a double boiler until it is liquefied. Pour the wax into the bucket (Figure 6–85) and roll the wax on the inside of the bucket until it completely covers the inside surface area of the bucket. Pour out the excess wax. (Figure 6–86)

NOTE: Both of these steps must be done *before* filling the bucket with water.

Figure 6–84

Figure 6–85

Figure 6–86

To leave the bucket natural:

Either submerge the bucket in water or else pour water into the bucket. (Figure 6–87) Then let the bucket sit for twenty-four hours. The bucket may leak a little at first, but soon the wood and flagging will swell together, sealing the bucket so that it will hold liquid. (Be aware that as you practice making buckets, you will eventually be able to make buckets that will not leak at all.)

Figure 6–87

NOTE: If the bucket is kept natural you must keep water in it at all times so that it will not dry out and fall apart.

You now have a completed Primitive Camp Bucket!!

Chapter 7
How To Care For Your Bucket

To insure long life of your bucket, I recommend you take the following steps:

—Store the bucket away from heat and sunlight. This will prevent excessive drying.

—Do not allow water to freeze in the bucket. It could break the hoops and spread the staves apart, causing leakage.

—Before drinking from the bucket, be aware that wood is a porous material and can harbor bacteria.

If you finish (oil and wax) the bucket:

—After a day's use, empty the bucket and let it dry. This insures that the water will not soak excessively into the wood.

—If you are using your bucket daily, recoat the outside with linseed oil or mineral oil and the inside with hot liquid wax every six months. This prevents excessive weathering of the wood.

—If you want to use the bucket for warm water, start with cold water and then stir hot water into the cold water. This will prevent the creation of a "hot spot" in the bucket. It is best not to use water over 120° F—the warmer the water, the more wax will melt off the sides which will cause the bucket to lose some of its sealant.

—You may notice a white film or spots on the inside of the container. This is caused from the wax and is not harmful. Use a hair dryer turned on high to blow hot air onto the film, and it will turn clear. Wipe off any excess wax.

— Before drinking from the bucket, be aware that wax, like wood, is a porous material and can harbor bacteria.

If you leave the bucket natural:

—Leave water in the bucket at all times so that the joints will remain swelled and the bucket will be able to hold water.

Chapter 8
Image Gallery

The following pages contain photographs of other buckets I have made. I hope they will give you some inspiration for other projects.

Old Oaken Bucket – This hefty bucket measures 11" x 11" x 13", and was essentially the "five-gallon bucket" of its day. It makes an excellent blacksmithing pail or well bucket because it is so durable. It requires larger staves made of oak, but otherwise the instructions would be very similar to those of the Primitive Camp Bucket.

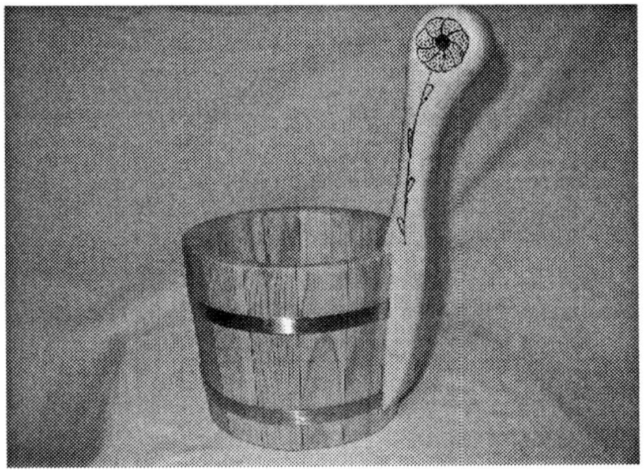

Danish Sauna Bucket – As the name suggests, this bucket is used in saunas for pouring water on hot rocks and the body. It is essentially a small bucket with a carved handle attached to the side.

Piggin – This was a farm container that doubled as a small scoop. It would have been used for general farm duties, such as milking the cows or scattering grain for the chickens. Although it looks somewhat like the Danish Sauna Bucket, its handle is actually an extra-long stave that is built into the bucket along with the other staves. Also, the piggin is built upside down, i.e. the head is added to the wide end of the bucket instead of the narrow end.

Katch-Lid Bucket – also known as a *firkin*, this French-style bucket is also built upside down, adding a solid split-ash bail fastened on with 1/2" hardwood pegs and a lid that has its own unique, H-shaped lift-off handle. The lid helps to keep debris out of the bucket, and the upside-down construction makes the bucket more difficult to tip over.

Conestoga Grease Bucket – This bucket hung on the back of Conestoga wagons and held lard and tar for lubricating the axles of the wagon. The bail is made of leather, and attaches in such a way that the lid cannot come off. This bucket includes a removable paddle for greasing the wheels; the lid fits over the paddle and holds it in place. This is another upside down bucket with only a slight pitch.

Dipper – The smallest coopered item I have made to date is a wooden dipper, which takes the concept of the Danish Sauna Bucket to more extreme proportions. I have made dippers that hold as little as eight ounces of water.

Water Tower – The largest coopered item I have made is this wooden water tower, which has been erected at The Littleton Historic Museum in Littleton, Colorado. It holds 1100 gallons and is 5' high and 7' across.

Bibliography

There is very little written information about cooperage, especially white coopering. Much of what I learned about coopering I learned from experience and observation. However, there is one book that is an excellent comprehensive guide to coopering that I have used as a textbook and reference guide over my years as a cooper. That book is ***The Cooper and His Trade*** by Kenneth Kilby (Fresno, CA: Linden Publishing Company, 1971, 192 pp.). It can be special ordered from your local bookstore or ordered direct from:

Linden Publishing Company
2006 South Mary
Fresno CA 93721 USA
www.lindenpub.com

I highly recommend this book to every aspiring cooper.

In addition, I have come across some rare print and video sources that will also be helpful to the would-be cooper. You should be able to obtain most of these sources via interlibrary loan, unless otherwise noted.

Booklets/articles:

Kilby, Kenneth. *The Village Cooper*. Aylesbury, Bucks, UK: Shire Publications Ltd., 1977. 32 pp. (This booklet is out of print and essentially summarizes the information found in *The Cooper and His Trade*. A good primer for the larger book.)

Langsner, Drew. "Making Wooden Buckets: White Coopering, the Swiss Way." *Fine Woodworking,* May/June 1983, p. 73–78.

Raiselis, Ron. "Tight Coopering: How Fire and Smoke Make a Watertight Cask." *Fine Woodworking,* January/February 1987, p. 40–45.

Brunson, Laurie. "Butter Churns." *Foxfire 3.* Anchor Publications, 1975, p. 369–397. (Note: page 381 of this article gives instructions for making a croze.)

Videos:

The Cooper's Craft: The Art of Colonial Barrel Making. Colonial Williamsburg Foundation, 1967, 29 minutes. Available on VHS from The Colonial Williamsburg Foundation, PO Box 1776, Williamsburg, VA 23187-1776.

Swiss Cooperage with Rudy Koehler. Country Workshops, 1992, 60 minutes. Available on VHS and DVD from:

Cambium Books
57 Stony Hill Rd
Bethel CT 06801 USA
www.cambiumbooks.com

or from Country Workshops (see information below).

Resources:

Country Workshops is a not-for-profit organization based in rural North Carolina. This organization specializes in personalized instruction in lost crafts such as spoon and bowl carving, chairmaking, and Swiss cooperage. Alongside the above-mentioned Swiss cooperage video, they offer the following resources for sale:

—Chairmaker's Froe…features a 12-inch blade; this is the type I recommend for stave cutting

—Shaving Mule (or Shaving Horse)…made of softwood or hardwood

Call them for details. You can request a free catalog from them at:

Country Workshops
990 Black Pine Ridge Rd
Marshall NC 28753
(828) 656-2280 (open 9 am to 6 pm Eastern Time)

or visit their website at **www.countryworkshops.org**.

About the Author

James Gaster is the owner and proprietor of Beaver Buckets, a cooperage in rural Nebraska that specializes in wooden buckets, butter churns, washbasins, and similar items. Mr. Gaster's handcrafted items can be seen in the Smithsonian Institute, the White House Visitors Center, the films *Seabiscuit* (2003), *Master and Commander: The Far Side of the World* (2003), and *The Alamo* (2004), the Lotte World Adventure amusement park in Seoul, Korea, and in numerous living history museums throughout the United States.

You can visit Beaver Buckets' online catalog of handcrafted buckets, butter churns, washbasins, casks, yokes, and other wood-crafted items at

www.beaverbuckets.com

To order additional copies of

How to Make a

COOPERED WOODEN BUCKET

Please visit our web site at
www.pleasantword.com

Also available at: www.amazon.com

Printed in the United States
20068LVS00004B/17